Other Books by the Author

(Available on Amazon.com)

White Godfather – Black Godchildren

Stepping Stones: My Tao in the Stream of Universal Consciousness

Pinned Butterflies: The Self-Captivity of White People

"Prediction is very difficult, especially if it's about the future."
(Niels Bohr)

"The future belongs to those who believe in the beauty of their dreams."
(Eleanor Roosevelt)

Generation ALPHA: Their World and Beyond
By Malcolm Scott MacPhater

Published June 16, 2021
Copyright Malcolm Scott MacPhater

(Cover photo credit: Julian Steenbergen – on Unsplash.com)

Contents

- DEDICATION 9
- INTRODUCTION 19
- CHAPTER ONE – WILL HATE KILL THE FUTURE?...........................23
- CHAPTER TWO – RED SKY IN MORNING 27
- CHAPTER FOUR – LONGEVITY, A NEW CLASS DIVISION 39
- CHAPTER FIVE – COMPLEXITY & FRAGILITY 45
- CHAPTER SIX – FUTURE LOVE, BIRTH AND DEATH 49
- CHAPTER SEVEN–ROUND, ROUND, GET AROUND 53
- CHAPTER EIGHT – WORK & LEISURE 57
- CHAPTER NINE– SPACE IS THE PLACE...65
- CHAPTER TEN – END OF RELIGION..69
- EPILOGUE..73

DEDICATION

I wish to dedicate this book to all those who dream of a better world. It is said, words have power. Dreams are the fuel of that power.

PREFACE

I suppose I have always been a closet futurist. What is a futurist? Simply put, a futurist looks around in the present world, considers the pace of past changes in our world, and then conjectures what the future world may look like. That "future" could be ten years out, 50, 100, or 200….as far along the "arrow of time" as one wishes to conjecture. Cosmologists look into the future to what may be the end of the universe, but for purposes of this book, let's try not to peer further than a couple of centuries!

As an example of a "futurist," take someone who lived during the Wright brothers' first flight at Kitty Hawk, South Carolina, who may have excitedly speculated about large "air machines" one day carrying people and things. By definition, that person would have been a "futurist." However, they may well have not even been able to consider when, or how soon, such large aircraft might carry people and things because there was little technological precedent to mark the pace of change in their world. From our vantage point today, however, we can look back over the last one-hundred years of changes and see that the pace of change has accelerated, and in many areas, the changes are astonishingly exponential.

There is no better example of the acceleration of such change than cell phone technology. I clearly remember the first cell phone I bought for my (late) wife, around 1984, at the cost of a few thousand dollars (and we think the new smartphones are expensive!). She carried it in her huge handbag, and I used to joke with her saying if anyone ever tried to attack her, she should swing that purse at them, for the weight of that phone would surely knock someone into tomorrow! Today our latest smartphones are millions of times more powerful than the guidance computers of Apollo 11 and may soon make laptops practically obsolete! Looking into the future, smartphones may one day be supplanted by mere chips embedded under the skin of our wrist, and a ritual for preteen's, while also getting their ears pierced (for "speaker" earrings), might be having their new "phone" implanted at the local "mall," (or whatever it may be called by then).

I sometimes forget my age, which by usual standards doesn't place me in the "oldest old" category yet, but I can laughingly say I am older than the People's Republic of China and many other nations. I was a country boy living with my family on a ranch in the Texas Hill Country when the Soviets launched Sputnik, the first human-made satellite to orbit Earth. We saw it passing over us one dark summer night as we sat in the yard sipping iced tea, trying to stay cool while slapping mosquitoes and taking in the hoots of our local great horned owl residents. Between sips, we would let our eyes wander upward, gazing at the dark sky filled with stars and galaxies, doing our best to catch a shooting star. My mother was, I suppose, the one who inspired my future "futurism," as she would often muse on those evenings of family togetherness, about what "might be" or "could be" or "what if." So I credit her with planting the seed in me, to wonder about what the future holds.

However, only recently have I began to think more and more about "the future of the world," mainly because I was blessed four years ago with a god-granddaughter, followed by her beautiful sister two years later (my wife and I were childless). *They are Generation Alphas*. So obviously, I wonder and am very concerned about the world they will grow up in and live out their lives; and, since we are trying to look out two centuries, what will the world be like for their successive progeny? My father's father knew the days when horses and buggies were the primary means of land travel, except for the railroads. He used regale me with tales of his "adventures" in those days, and I would listen with rapt attention as he would imitate the sound of a panther stalking him in the cedar brakes of the Texas Hill Country. As time passed, my grandparents lived to see Neil Armstrong walk on the Moon, and even more fantastic human accomplishments. From horse walks to Moonwalks is a stretch, to say the least, basically less than three-fourths of a century! I now am 75, and what I have seen transpire informs much of my thinking as I write this book.

My god-granddaughters were brought into this realm by my eldest god-daughter and husband, who are now successful black millennials living in

NYC. Why do I even mention that they are "Black"? Mainly because I am white, and this book must and will speak to aspects of race and racism now and what "it" might be in the future. Will America "have the conversation" and do something from that, or not? I have addressed this topic thoroughly in my book,

Pinned Butterflies: The Self-Captivity of White People.

The first god-granddaughter, would you believe, was born on my birthday, April 25! So my perspective looking forward into the world that my "birthday girl" and her siblings (we expect more!) will grow up in, and likely contribute her DNA to the next generation, is one of great concern. Can I do much, if anything, to change that world if it needs "fixing"? The answer, of course, is "no," but what I hope to do is share my insights and observations with her parents and all the parents of Alphas who may read this book. Maybe then they can make some decisions ahead to better prepare their Alphas for the future they will find themselves living. Of course, I also hope what I have to say can stimulate the thinking of other readers, movers and shakers and policymakers, to help move the "future change curve" closer in parallel to the "arc of justice," although both are long and slow to bend.

While no one can see into the future, we can observe what is present and then conjecture about what the future might be. However, our perceptions of what the future may "look like" depends on both objective and subjective variables of the world around us and who we individually are. How we individually see the world and its future largely derives from our minds since we can only create the future within our minds. However, Buddha teaches that "all is Mind." I believe that, and believe that what we think taps into the potential of the Universal Consciousness, so thoughts (and prayers) may manifest as our reality.

As I set about writing this book, the complexity of peering into the future came clear immediately. The more I looked into futurism literature, the more I saw that the crystal ball "revealing" the future is an infinitely layered and complex onion. Peel back one layer in a tiny area, and a dozen more layers can be seen beneath. Our human-made world each day expands in complexity, and the decisions we make and act upon also daily impact the natural world. Impinging on the overall complexities are the "unknowns" and the timing of "mega-unknowns." For instance, humankind's effect on the world climate is already causing severe melting of Artic ice and gargantuan icebergs to break off the Antarctic ice shelf, thus affecting the rise of sea levels and the temperature of ocean currents around the globe. While scientists may predict with some accuracy when some some events may occur, say within ten years, they cannot say for sure, nor can they communicate with specificity how much the sea levels or temperature will rise. One thing is sure though, the seas will rise, lower elevation coastlines throughout the world will be lost to the great ocean, and people inhabiting those areas will be displaced, further exacerbating crowding and competition for resources, particularly food.

Moreover, it is already evident that vast areas of farming and cultivation are being wiped out yearly. Whether it is the burning of forests in Brazil or other struggling countries, to the fires, floods, and droughts in this country – the breadbasket of the world – or plagues of voracious or harmful insects such as the locusts in East Africa recently, all contribute to the reduction in the world's food supply.

Deserts are expanding every year, in Africa and here in North America. Likewise, severe droughts, overuse, and misuse of water in our country's West and Southwest have ruined orchards and farmlands. We are witnessing similar freshwater depletion in many areas of the world that threatens the survival of humans and animals. Since we cannot expect Mother Nature to respond to our wishes suddenly, the question is what we can do now or soon that will help arrest the water crisis, or provide alternative sources of water; or, can we find ways using less water to grow food for the world? (One thing we could all do is reduce our consumption of meats: a tremendous amount of land and water goes into the production of cattle and chickens.)

Then there is the vital issue of maintaining the air we breathe - that all life breathes - to assure a future for us, the fauna, and flora. How do we offset the loss of rainforests, the loss of coastal kelp forests? These are huge questions pregnant with importance to our very survival.

However, most important is that these kinds of events will happen, and so it is important for us to speculate on what the spectrum of effects may be on societies and civilization. However, such speculation can only be pure guesses. The plethora of technical, social, and geopolitical reactions to such speculation can only be addressed based on how people and cultures have reacted in the past. Human nature being what it is, fraught with subjectivity and bias, makes even such projections very tenuous. However, since our present everyday lives are conducted in a whirlwind of "probabilities," with nothing a sure thing, "except death and taxes," we should not have trepidations about examining what the future may be. How else might we avoid or at least mitigate what would otherwise be significant calamities a decade, or ten or twenty decades, from now? Dreamers and planners laid the foundations of all civilizations. *Now futurists are called to predict the effects of humankind's actions, so policymakers and planners can mitigate decisions that are net negative.*

Nothing we humans do has only a single effect. Everything we do in or to our world, technical or natural, has at least one "other" product, often negative. Looking into the future, we must be wary of what we do, for even well-meaning actions can have drastic negative consequences. Examples of such projects negatively impacting the natural environment include the Suez Canal and projects long ago that connected the Great Lakes, allowing invasive species to move into waters where they had never been before. Another good example here in the USA is how farmers used fertilizers in the Mississippi Valley, which flushed down that mighty river into the Gulf, has killed off a great swath of coastal undersea life.

What our future world holds for us and our progeny is not merely conveniences such as faster transportation. With all technical and natural shifts, there arise social and psychological responses. Take AI (artificial intelligence), for example, and more specifically the AI of human-like robots. Even as I write this, small factories are churning out female and male robots, life-sized, designed to a buyer's preferences. Whereas today there are advanced versions of "sex toys," in the future assuredly, these androids ("early humanoids") will play a much greater role in the lives of many (who can afford to purchase them). Androids will, in the not distant future, supplant human partnering for those men and women who are unable to find a partner or find one that they can remain happy with. Much later will come the moral and ethical questions of whether humans and androids can marry, can androids have legal rights, can an android own things, can one's android lover be sold/reconditioned/ (terminated?) and replaced with a younger/hotter one?), and so on. At some point, when robots' intelligence is far greater than the human brain, we will also have to ask, do they have souls? These sorts of questions were touched upon in the film *AI ARTIFICIAL INTELLIGENCE (2001)*. Moreover, given that one may surmise that androids could "live on" forever, long after their mates, then what? Since they would never age, someone might wind up marrying the humanoid lover (wife?) of their own grandfather!

While robots designed for other areas of human interaction may not need to be human-like anatomically, no doubt we will see robots in various capacities such as "big box" stores (where presently it is not easy to even find a real "salesperson" to ask a question). In time, any job that can be filled with a robot will be. That is the future we must prepare for!

As I embarked on writing this book, one thought kept returning: how can we think about the far future when we do not know if tomorrow is even given to us? Moreover, if one is a person of color, the promise of another day in this world is even more tenuous. Every day men and women and children of color fall victim to gunfire in numbers far greater than their relative populations, and almost daily are the reports of a person of color, usually, a young black man, being killed by cops, not always by bullets.

Earlier this year, we saw an attack on the United States Capitol. Would we, in our wildest speculations, have ever thought such an event would occur? I certainly would not have. There is a high probability that 2022 and 2024 will again see such horrific events! In this period of our history, we are indeed living in very troubled times; yet, from what I recall over my 75 years on Earth, particularly since the early '60's when I was old enough to be aware of major events, every period could have been labeled as "troubled," yet somehow we came through each. My expectation is that this too shall pass.

Out of each period, I choose to believe that we came away having gained or learned some positive things, which are in our tool kit to deal with coming years. So despite the gun violence and social/political turmoil we are witnessing in 2021, *I believe there is a growing sentiment that collectively, we must find a better way.* There is no alternative if one thinks through the matter: if we cannot find a way to live together in relative peace and fairness, such matters as climate change may be moot. If we cannot have a robust, productive, and reasonably fair society, and continue to stumble along a path of pending social collapse, then really does climate change matter to us? (The answer is, "Of course it must.")

So there are good reasons for us to lean forward, looking into "what could be" so that we may nudge our destiny toward the positive. However, social norms, diverse attitudes about ethics, and mainly, the all too obvious and insidious perversion of conspiracy theories that abound today and no doubt will propagate like weeds into our future....all will be factored in the mix of how societies evolve. When today we have those who promulgate nonsense such as "viruses do not exist," one can only shake one's head at what such ignorance portends for the future. All this should make the rest of us even more committed to positive change, particularly toward improving our educational systems. Until history is presented factually, and until our children are taught the importance of critical analysis of information, the outlook for the future is dim.

INTRODUCTION

The idea for ***Generation Alpha: Their World & Beyond*** germinated from the author's love for his two god-granddaughters. He, being at the leading edge of the Boomer generation, is greatly concerned about the world in which the two little girls will grow up. And what about the future worlds awaiting their children, and their children's children? Based on what we can see around us today, one cannot help but have grave concern for the future of Gen Alpha and generations beyond.

Like all Boomers alive today, the author has had a seat to several chapters of the history of the United States and the world. Born in April 1946, he recalls when radios and television used vacuum tubes and vividly recollected his father turning on their first black and white TV in the living room that doubled as the author's bedroom in 1952. The family enjoyed watching the programs they could pick up out of San Antonio, some 100 miles to the east. His father installed a 50-feet tall antenna by the old ranch house, which often swayed in the pressure wave following F-100 jet fighters flying tree-top level missions down the river canyon where the ranch was situated. His dad also bought a "rabbit ears" antenna that he placed on top of the TV cabinet, and took some tin foil and attached it to the ears, which created a fine-tuning capability; i.e., they could play with and jiggle the foil to find the best reception! Malcolm Scott MacPhater (the author's pen name) also remembers making his first radio, circa 1960, one he totally built from "scratch," that used a crystal diode and Army surplus earphones. Some years in his future then he would become an electrical engineer.

The fact is that over the nearly eight decades since the first Boomers arrived, science and technology have changed our world in countless ways that were wholly unpredictable. There were then, as probably across human history, those who dreamed about what the future might be. Today humankind stands on the threshold, perhaps better called the tipping point, of potential shifts in our world that offer two doors: one may lead to the "salvation" of we Homo sapiens at best, the other to cataclysmic collapse, at worst. Which will we choose?

MacPhater dons his wizard hat for this book and summons his appreciation for history and his curiosity of scientific, technological, and societal matters to peer deeply into the proverbial crystal ball. MacPhater is a retired electrical engineer who participated in designing and constructing all types of facilities and the "built environment" over a fifty-year career, from top-secret government projects to the corner taco fast-food "joint." He served as a junior officer in the United States Air Force during the Vietnam era where he worked with intercontinental ballistic missiles, and often managed people and projects over his career. Since retirement, he has written several non-fiction books, but this one he states has undoubtedly been the most challenging.

Many of the subjects he addresses or touches upon herein are ripe with unknown, incalculable potential for shaping the future world. For instance, current particle physics research on "muons" appears to be leading to the discovery of previously unknown quantum virtual particles, which could rewrite the "standard model" of quantum mechanics. As with all things that human beings have "discovered" over centuries and millennia, there always seems to be more to find just over the horizon.

As the author notes, taking on the self-challenge to write such a book as this proved daunting but certainly gave him free rein to let his mind roam across the plains of potentials. He has allowed his innate curiosity about all things to propel him through future vistas and has enjoyed this romp of what the future holds for his god-granddaughters and their offspring generations. In spite of the challenges awaiting in the future, he remains optimistic.

MacPhater encourages readers to seriously pursue paths that extend their thinking into the near and far futures, whether across the panorama, or focused on specific aspects. He believes the world needs what futurists can bring to the planning and policy tables if we want to make this world better, not worse. Presently civilization has entered an era that, as mentioned earlier, can be truly characterized as a "tipping point": only by putting reason over selfishness will we avoid a global domino effect leading to a full collapse of our natural world and human society.

CHAPTER ONE – WILL HATE KILL THE FUTURE?

One may well ask, why would I begin this book with a lead chapter asking this question? As I mentioned in the Preface, my two young god-granddaughters are Black. Given that, still in 2021, we see a rising of overt racism and xenophobia, the likes of which we haven't seen since the Jim Crow times and German Nazi era, the question for me is as important, if not more so, than any other about the future.

This type of hate and evil has its seeds in the Atlantic slave trade. I do not wish to delve into a detailed discussion or argue with those I know would push back on this statement. Still, the fact is that it was chattel slavery, the dehumanizing of Africans brought to the Americas, then the Civil War, post-antebellum years, and KKK leading to the Jim Crow decades that got "us" here today. But racism is just one of the many ugly faces of hating "others." The recent rise in vicious attacks and murders of Asian and Pacific Islander people shows another face. The continuation of Palestinians held in apartheid by the Zionist government in control of Israel is another. Of course, I cannot forget the hate attacks on LGBTQ people, especially murders of transwomen, are higher than ever. We must look deeply at the root causes of such hate, and then as futurists, look at how future conditions and circumstances manifesting will affect these causes.

We could point to many antecedents of racism and xenophobia, but I believe segregation of people of different ethnicities and hues of skin color is where it begins. When people blend in over time, most perceived differences diminish if not disappear. I am not referring to "desegregation," as school systems dealt with in years past, for by in large, students still self-segregated socially due to the overwhelming weight of systemic racism. I am referring to situations today mostly occurring in "micro-social" groups in the USA and Europe, where there is true social mixing, from dating to marriage to like-minded social groups of persons in cross- or bi-racial relationships, including across religions. (I dislike using the term "race," but our English language is deficient in this regard.) This phenomenon of mixing, whether in-person, online, casual, or intimate, is increasingly permeating the matrix of diverse human values. I remember when no family would admit they had a child who was gay or lesbian, and then in time, even some of the most conservative public figures opened up, accepting their LGBTQ offspring publicly. When this occurs, and people have to deal with one another as persons, not "this or that," then a common understanding at a very human level can occur. Hate is a virus that spreads, but I believe that love, the antidote to hate, is more powerful over time. Love begets justice, but as was said by Dr. King, "the arc of the moral universe is long….;" we must be patient.

So then the question arises, in the future, how much of this "dissolving" of differences will occur? Given the myriad influences that will shape and determine how this plays out and how such relationships evolve in various parts of a dynamic world cultural ecology, we can only firmly hold onto our futurist hats and venture a few informed guesses, tinted by our pessimism, or optimism. I lean toward the latter, yet well aware that there will be racists "holdouts" who cling to their premise of separation of the "races."

Hate often derives from fear, fear of the "other," fear of change, and so on. Fear mixed with the history of guns in the US, and misconstruing of the 2nd Amendment to the Constitution, have brought us the world we live in today, with a mass shooting, or shootings, occurring daily. I grew up around and with guns….hunting rifles such as 0.22 caliber single-shot and Winchester .30-30 lever-action rifles, and the high-power .30-06 caliber, and shotguns. I was raised in the Hill Country of southwest Texas, sheep, goat and cattle country, and every rancher and household had guns, mostly for hunting, or to protect livestock from predators; there was very little fear back then of a home invasion (I doubt if the term was even coined back then).

But we have become a gun culture, and guns are purchased and owned by whites disproportionately compared to people of color. Over the last year white people have bought millions of guns! What we have is a public health crisis, a mental crisis of gun addiction. As I write this, the Texas Legislature has sent a "permit-less" gun carry bill to the Governor! And this is supposed to make us safer? And, a Federal judge has struck down California's ban on

AR-15 assault rifles. My god! (And, believe it or not, it is against the law in Texas for even an elderly person to carry a club for self-defense!)

In my world view, hate will always be endemic to the human race, in one form or another. It results from fears and competition tensions on many levels, and race or ethnicity "othering" is always an easy path for those who will not look at history, nor themselves, with clear eyes. Since this book is about futurism, I draw from this that futurists also must see ahead using the light of moral conviction. But this translates into a belief, at least a deep hope that humankind will find that everyone should set aside petty race hatreds and focus on the growth and mutual welfare of all.

This could be said to be "pie in the sky" based on historical and current events. Yet as time goes by, it appears there is a growing consensus that "enough is enough" when it comes to all aspects of hate and its manifestations, so if enough people of all persuasions believe this, collectively, we can incrementally direct our cultures along that moral arc. Yet, the same fundamental social, societal, and technical factors that will impact and impart momentum to our world along the decades and centuries ahead play right into how we feel about each other as to "race."

Indeed, the "blending" I mentioned above…interracial relationships that are more and more common, bringing more and more bi- or multi-racial children into the world, will increasingly bulwark the bridge between peoples of different hues or cultures. Nothing can stop this, for love always trumps hate. That has been proven time and again, especially during periods of war and conflict.

Also, it is very apparent today that the present Gen Z, and those generations behind them, are tuned to a world of social media and internet associations with those of their age group who often do not look like the kids next door. Through social media, people see that people are much like themselves around the world. The internet is both a window and a mirror. This cross-cultural socialization is also very apparent in many of the newer suburbs of America, including the towns of Collin County, Texas, where I presently reside. Right in the complex where I live, there are young people from all over the world, who came here to study or work in high-tech employment, or the healthcare field. It is very refreshing to see this phenomenon.

However, there are countervailing factors that will tend to retard these new self-identities. One is the growing differences in economic "classes." It is apparent that with the impact of technology on jobs, more and more "blue-collar" working people will be squeezed out of the better-paying jobs due to the rise in the use of AI and robots. And, even below them economically is a growing class of new immigrants hungry for a piece of America. Yet looming far above all these people - the poor, the shrinking middle-class, and small independent businesses - is the super-rich class, who over the last decades have seen their wealth multiply. The divide between the multi-billionaires and "the others" has widened exponentially. Typically when people are squeezed to compete for more within their classes, tensions build, often along the lines of "othering." So we must continue to be diligent in heading off such battle lines in the future.

CHAPTER TWO – RED SKY IN MORNING

Many of us are at least vaguely familiar with the adage, "red sky in morning, sailor takes warning." I am not a sailor, but I feel this theme connotes a sense of the ominous that climate change portends for the future world of Gen Alpha and their progeny. Besides the pandemic that is now raging as I write this, there are no other threats to the continuance of global civilization that are so obviously imminent.

That's not to say there aren't other phenomena that could dramatically shift or even end human culture and much of nature on all surfaces of our planet. A collision with one or more giant asteroids certainly has that potential for doom, and such objects hurtling through space are out there, and even though we currently track many, many more (and who knows how many?) go unseen. Such an asteroid collided with Earth about 65-million years ago. Six miles across and perhaps traveling at 20 miles per second, the kinetic energy released upon its collision changed the course of life on Earth. Another such collision would undoubtedly bring an end to life as we know it. Or, what about a flipping of the magnetic poles? Or, a massive burst of solar flares that cooks all electrical systems here in certain expanses of the globe…such as the USA. Since we do not like to dwell on such catastrophes…in fact, our minds find it difficult to even accept the possibility of such an occurrence…for now we will set aside consideration of such events, and "worry" about those we may be able to do something about!

Today's babies, toddlers, and third graders will grow up in a world of increasing extreme weather events, rising seas displacing people and remaking coastlines, and the gross effects of climate change on the natural world. Scenes of nature that were enjoyed by humanity during my lifetime will become no more than images of extinction, much like the sky filled with carrier pigeons or the vast moving herds of buffalo of the late 1800's…that white men decimated. When I was a boy growing up on a ranch in the early '50's, nature was no longer pristine in that area, but compared to what it is like today, it was a Garden of Eden! Today, due to how the land and rivers have been "managed," and due to the incessant movement of people…mostly white people…away from major Texas cities into the area where I was raised, the Garden has long ago vanished. So it goes in many areas of the world.

Perhaps genetic engineering and other biological projects will save…or clone…disappearing species, but nature will have been altered either way. While great efforts to sustain certain species by moving endangered animals by plane and boat to conservation game preserves, many to be seen today around Texas and in other parts of the world, the natural habitats are irretrievably gone for these animals. Overlaying this phenomenon is the commercial business of capturing and moving animals of all sorts from their natural runs to become pets thousands of miles away from their place of birth. Most of these animals that survive wind up being let loose to forage, breed and die, and in the process disturb and even threaten the continuation of indigenous species. A case in point is the thousands of pythons in the Florida Everglades.

Here in Texas, we recently experienced cold as never before (Winter Storm Uri), which, coupled with a failure of the electric power grid in the State, resulted in a catastrophe. Millions of shallow-water ocean fish along the Gulf Coast were killed by the freeze! Being a retired engineer myself, I know that engineers and architects typically design for "100-year events," but such benchmarks quite likely will need to be rethought and modified. Floods, droughts, wildfires burning millions of acres of foliage, wildlife and towns, tornadoes and hurricanes coming in droves….this is climate change, and it will only get worse. Even if we can curtail the gases and pollution we continue to pump into the air we breathe, we cannot expect much improvement until one day, far in the future, we might be able to help Mother Nature heal itself. If we can find it within ourselves to be better stewards of our world very soon, perhaps in time, Nature will begin to heal itself some.

Looking at these effects globally, assuredly, we will see the rising sea and increased ferocity of typhoons and hurricanes drive fragile coastline cultures inland. This will cause severe competition for food and resources in rural and urban areas, resulting in continual conflict. The United States government security agencies have already identified these factors as sources of potential conflict and threats in the future, beginning very soon.

Here in the USA, the coastline will also be severely impacted by the rising tide line. The encroachment of the sea and sub-surface saltwater is already a big problem in some areas. The impact of increasingly frequent hurricanes is exacerbated by the rising sea, as each storm pushes water further and further into areas previously protected by levees, walls, and pumping systems. We saw what happened in NOLA when Katrina hit. I lived through Harvey in Houston, and one day even a worse storm could directly hit Houston, "God forbid"!

Sure, hurricanes and tornadoes have always been phenomena that threaten us. The apocalyptic hurricane that struck Galveston in 1900, or the F-5 tornado that wiped out the small town of Rocksprings, Texas in 1927, where my father later attended high school, occurred before the term "climate change" was even thought of. But the clearly increasing frequency and severity of storms has been directly linked to shifts in the sea currents, ocean temperatures, and global weather patterns brought on by climate change.

While storms are like exclamation points of our disturbance of Nature, rising seas are gradual and yet inevitable. However, embedded in the complexity of our climate, changes are happening that are turning land that has been cultivated seasonally for thousands of years, into desert. Also, kelp forests along the coastlines that contribute a high percentage of oxygen to our air, are threatened, and many are dying off. Tropical forests are being destroyed by humans, as in Brazil, while wildfires consume millions of acres of flora in the US West, increasing every year.

So one must rightly ask, how will humans flourish, much less survive, as time goes by? Do we fall back on the adage, "necessity is the mother of invention"? More likely, it will be a combination of invention and adaptation, coupled with highly undesirable "fallout." Fallout meaning, in this case, large segments of the world's population "left behind," deprived of inventions and adaptations in which the more affluent societies will seek their livelihoods. Think of cities, high-tech food growing farms, water generating plants…..all under domes, surrounded by vast areas inhabited by exposed and struggling "others." Not a pretty picture, is it? I can think of a few movies based on such scenarios.

The acceleration of climate changes worldwide will undoubtedly result in forced changes in how people survive, such as populations moving inland; but, developed urban societies will also adapt by creating ways and means of mitigating the effects of changes in the local climates and weather extremes. An excellent example of that is occurring right now here in Texas, where the aftermath of Uri taught us that our energy infrastructure requires significant "hardening" to extreme cold. We will soon also find that similar measures will be needed to deal with extreme heat events, as we should expect summer temperatures to exceed past highs immensely and for many days longer. We have seen what such high heat events can do, especially in northern cities not equipped with high-capacity air-conditioning systems within residences. Those kinds of incidents likely will seem like a cakewalk in the future, so we must prepare.

Where there is a need, there is a future invention. As an engineer, I firmly believe in that. So I am optimistic that humans will develop creative solutions to dealing with extreme cold and heat. A straightforward way is to improve insulation systems significantly, use ancient methods as employed in New Mexico and Arizona areas for decades, and build underground, perhaps one day, even undersea.

But in years and decades to come, technology will offer solutions, at first too expensive for most households, that will provide more excellent and more efficient protection of our living environs. Technology will also rapidly drive down the cost of new inventions and provide for more manageable and wider deployment, as is already being done with 3-D printing design and construction. One can only pray that such technology will promote wide distribution across the swaths of impoverished areas here in the US, giving everyone a decent habitat - and also in "Third World" countries. We have seen how quickly and widespread cell phones have become in those areas.

Also, in many places, we see the growing use of photo-voltaic generation and other "First World" technologies, so future inventions may spread quickly to protect fragile human beings everywhere from weather extremes.

As we are growing up, we naturally take our ambient conditions for granted and think little of it. When I was a kid growing up in the country, our family lived on the edge of poverty. I thought that was normal. It was only when I brought a school friend home for a sleepover and he seriously asked if our house was the barn that I began to see things differently. So children of today, Gen Alpha, will generally accept the circumstances they will grow up in, to the point of "awakening" to the meanings of and reasons for their surrounding world. However, the Millennials and Gen Y are now saddled with reacting to the new climate, weather, and effect on societies. Will these two generations be able to correct the tuning of the engine of climate, to save it for Gen Alpha?

CHAPTER THREE – $E=MC^2$

Albert Einstein's famous equation is known to any advanced STEM student: Energy = (Mass) x (Speed of Light) Squared. The "E" – energy – raises the question of how future societies worldwide will produce energy and convert energy into power. An attendant question also arises, will there be sufficient energy and will it be equitably shared? Energy, especially electrical energy, has always fascinated me….which led to my becoming an electrical engineer who enjoyed a fifty-year career designing and building electrical power systems. During my career, I was fortunate to be deeply involved in various electrical power production projects, from sizeable gas-fired boiler/turbine generators to gas-turbine engine generators, to large solar systems. Over my career, I witnessed extensive changes in technology related to power production, its control and its use.

As I alluded to in the previous chapter, Texas discovered it has a severe problem with its power grid this past winter, as winter storm Uri blanketed the State. For the most part, this problem is solvable with current technology and proper management, with enough populist will and political capital exercised to affect the needed changes in the regulatory structure. The root cause for the failure of the grid has one leg in greed and the other in this "don't step on me" attitude of many Texans on the political right. It is rather odd that Texas has the largest wind generation and solar (photo-voltaic) capacity in the nation. Also, beginning a number of years ago, the State embraced "green architectural design" of new facilities. Personally, I was proud to participate as an engineering designer in the first decade of this revolution.

But let's do our best to look into the future of energy, not just in Texas, but around the world, for although there may not be electric lines connecting most nations, there is no aspect of human endeavor that does not have a "butterfly effect."

The fact is that energy consumption will continue to increase in all areas of the globe, whether because of the exponential addition of devices to the "internet of things" or the rapidly increasing industrial production in developing nations (and production growth of developed nations). Thus the question arises, how will this energy be generated in the coming decades? Today several technologies are in use, and in practice, all these technologies are intertwined, mainly via the "wires" that connect them ("the grid"), and connect users (households, businesses, factories, et cetera) to that system of generation and distribution of electric energy. Let's look at these briefly. Large generation stations that rely on coal or natural gas, still constitute a significant source of power in the US. Nuclear power – fission technology – plants built in the 1970's into the 1980's – also powers the grid, rain or shine. In the US and other developed nations, solar (of the photo-voltaic type) and wind turbines are also significant contributors, followed by various other "cogeneration" power technologies. However, these "green" sources can only produce power when the sun shines and the wind blows, so until adequate electric storage facilities are available, for some time we will continue to need the fossil fuel plants.

However, it has been known to climate scientists and environmentally knowledgeable policymakers for years that the old technologies that rely on fossil fuels are also significant heat injectors and pump millions of tons of particulates into the atmosphere annually, exacerbating climate change. Unfortunately, the need for the kilowatts generated by these plants around the clock, compared to green alternatives of solar or wind-dependent turbines, is real. Therefore solutions to get us past this conundrum presently are: reduce overall consumption via energy efficiency measure and incentives; use of energy storage systems such as large utility batteries; and, lastly, consideration of once again building nuclear fission plants, and also hoping for atomic fusion technology to finally prove itself and become commercially viable. Of course, anyone who has been around since the '50's knows that fission nuclear plants produce atomic waste, with the concomitant problem of how and where to store this radioactive waste. For decades, disposal and storage sites have been an environmental hazard and are still so very much today. However, fusion nuclear plants resolve that problem, but the technology of the basic fusion reactor is akin to harnessing the energy of the

Sun....not easily done. This technology has been researched and tested for a number of years, and current prototype reactors now hold great promise.

Looking into the future, it seems inevitable that fusion power will become viable on a large scale, hopefully within the next 10-15 years. We can also be assured of other developments that will result in a much greater efficient use of energy, several which exist today but await the economic "right time" to become part of the normal landscape. Many such developments will likely be devices/apparatus/cyber that give consumers the capability to generate and control their own power production and power-using devices that are incredibly efficient. One day we may see many homes and domiciles built that are not connected to the grid and equipped with their power plants, possibly micro-fusion reactor systems, or a still "un-invented" technology. We can also expect all occupancies will be so energy-efficient that their power consumption will be a mere fraction of a typical residence today. When Generation Alpha and their progeny no longer have to worry about paying a high "light bill," they will be able to spend or invest that equivalent income on many other benefits to their lives and lifestyles.

Eventually, the tens of thousands of miles of overhead electrical transmission lines and distribution lines will be removed, replaced as necessary by totally underground lines using super-cooled lines, and possibly even some other technology of transmitting power that is unknown today. The present method of power distribution that began with Thomas Edison is highly inefficient due to the power losses in the transmission conducting process. Overhead lines are also exposed to the weather and sabotage, so the eventual replacement of these lines with an underground, low-loss "super grid" will realize a much greater efficiency, reliability, and security for power networks. We should also expect radically different technologies in how networks "step down" voltage levels to the final outlet for appliances. This will result in doing away with the giant steel and copper transformers seen in substations and on power poles across the country, again improving power delivery efficiency. By vastly increasing electrical distribution efficiencies, relatively less power will have to be generated at central power plants.

Technological advances will also result in power users....machinery, air conditioning, lighting, et cetera....realizing exponential improvements in efficiency over today's world. Much of this technology will continue to result from space exploration developmental programs. This again reduces the total amount of power that must be generated to meet demands, reducing the heat and by-products of power production.

As I present elsewhere in the book, at some point in the future we can expect divisions of economic classes to be much more extreme than what we see today, which in itself is terrible. In India today, billionaires erect high-rise castles in areas where the masses live in filthy streets just outside the gates, much like the story of Prince Siddharta, the boy who became Buddha Shakyamuni. Guatama was kept from venturing outside his father's castle perimeter walls to prevent him from seeing the real world of poverty, hunger, suffering, and death. While statistics indicate that there is less hunger and poverty in the world today than say 50 years ago, we are looking at a future where there are strong reasons to believe this trend will be dramatically reversed if we do not stabilize the main predicate of food scarcity, climate change.

As I predicted elsewhere, given the fact that technology will tend to divide what we have classically called the "haves" from the "have-nots," we can expect there will be protected cities or large enclaves for the "haves," with the "have-nots" residing throughout what we now call rural America. This change in the social landscapes will likely take similar shapes in Europe and other societies. At that time, energy will become a highly protected resource, so we can expect those excluded from the "have" areas to be finding ways to obtain energy "illegally," much like Africans left out of the petroleum energy boom in Nigeria "steal" oil and gasoline from leaking pipelines. The schism in access to adequate energy resources in impoverished areas will lead to more significant conflicts in the future, especially in areas where the forests have been depleted by clear-cutting and cutting to make charcoal, et cetera. However, the technology exists even today to provide relatively low-cost and plentiful energy sources in such areas, but that depends on the humanity of the "haves." I have little reason to believe human nature will change that much over the decades and centuries, so I expect classism to continue to divide us Homo sapiens.

CHAPTER FOUR – LONGEVITY, A NEW CLASS DIVISION

Methuselah is said to have lived to be 969. We are entering an age when people may in fact live to be hundreds of years. Whether that ever happens, of course, depends on whether we stave off or conquer other major threats to our longevity, such as war and climate change. However, I believe it is highly likely many humans will live much, much longer, and that those we consider to be "old-timers" will be merely "whippersnappers" at some point in the future.

There is already a tension building between what science and technology can do to "modify" the human organism and even extend human life versus how equitable such "advances" can be. Science and technology are at the edge of an explosion in the realm of human biology, which will lead into regions that only "God" played in previously. However, there are many serious ethical issues that society must, right now, began to address head-on. We know ethics and values change over time, so future generations will be born into a world that we would find hard even to imagine in all ways, certainly ethically and morally. However, beginning even today, there is a substantial underlying issue of who really will have access to, and benefit from, such advances?

Given that today that only the rich and mega-rich have ready access to the "best" medical and health care, will these advances coming in the future only be available to the wealthiest? Or, will populist opinion drive medical care to finally be realized as a human right, and all people should have the same chance to the best in healthcare? Probably the richest will continue to have private access to the latest inventions that may save one from a particular disease, but at large it is also probable that the "average American" will have access, without cost being the limiting factor, to top-grade treatment, medicines and procedures.

As in all other areas of society, medicine and related treatments and therapies will generally include robots, humanoids, and AI. We can be sure that hospitals and nursing homes will eventually be staffed by humanoid robots who cannot spread diseases or be affected by viruses like COVID-19. Such robots will be so real that they will tirelessly attend to all the needs of residents and patients, providing even the touch and caring bedside manners that the best doctor or nurse today offers, and never tire of exchanging small talk with the patients.

We can expect medical clinics, "doc in a box" locations, to be primarily staffed by robots and very high-tech diagnostic apparatus. We also will see in-home diagnostic and treatment solutions become prevalent, to the point where the average family doctor will become obsolete or migrate into a new role yet to be defined. We have entered this new age of patient-doctor relationships already, with virtual doctor visits in common use, and an array of self-testing, diagnostic devices, and remote-monitoring readily available for purchase over the internet or at the corner drugstore.

Today we can clone mammals of all types. Likely a human being could be cloned (and may have already), and it is only laws and ethics that continue to prohibit such experiments so far. Dolly, the sheep, was the first major cloning success, and routinely other mammals are cloned for various reasons. As I write, research is well underway to attempt cloning the extinct wooly mammoth! No doubt, one day, DNA of one or more dinosaurs will be found that can facilitate its cloning, which might have to be done in a giant artificial "egg" of some sort. Whereas today there are a number of mechanized exhibits of dinosaurs, one day Jurassic Park may actually become a reality.

We cannot rule out that far into the future, for reasons perhaps we cannot fathom today, cloning of people will become common. Whether this might happen in 25 or 75 years is anyone's guess, but personally, I believe it will become accepted within the next 100 years.

Genome sequencing and gene therapies will soon be deployed against cancer and other diseases. Nanotechnology is/will be used to deliver drugs, heat, and light to specific places in the body, perform diagnostics, and treat wounds and other injuries. So in many ways, the future is knocking at our doors!

What the status of medical diagnosis and treatment will be in the future is intriguing to contemplate. Experiments conducted in space will undoubtedly find applications to the medical field, as we have benefited from for decades already. Here are just a few examples:

- Digital imaging breast biopsy system, developed from Hubble Space Telescope technology
- Tiny transmitters to monitor the fetus inside the womb
- Laser angioplasty, using fiber-optic catheters
- Forceps with fiber optics that let doctors measure the pressure applied to a baby's head during delivery
- Cool suit to lower body temperature in treatment of various conditions
- Voice-controlled wheelchairs
- Light-emitting diodes (LED) for help in brain cancer surgery
- Foam used to insulate space shuttle external tanks for less expensive, better molds for artificial arms and legs
- Programmable pacemakers
- Tools for cataract surgery

AI will multiply the power of human physicians by powers of 10, even 100 one day. Robotics already used for years in performing critical surgeries such as prostate operations will be commonly used for just about any type of operation or treatment. One day the staff in your doctor or dentist's office may be mostly robots, then one day humanoids! Children should find this quite entertaining....but by then, kids will accept such non-human presence as nothing but ordinary!

Gene editing tools that will make today's CRISPR (that got us the vaccines for covid) look like a Ford Model T does to us today! What does this mean? Will "natural selection" of our biological procreation process continue to occur within only the "have not" populations, whereas those in the "strato-classes" will be able to have their offspring "engineered" to their specifications?

Since, at the time of this writing, we are still in the midst of the covid pandemic of 2021, it is easy to imagine how tempting it will be for gene modifications to be "injected" into the animal kingdom as we attempt to modify nature to eradicate certain viruses and bacteria. But any such intervention in nature will certainly result in a reaction that we may regret.

I suspect that there will be a new field of study at some point in the future – and I have no name to give it – in which scientists use AI and access all the cyber data stored in "the cloud" about a particular human after they have been deceased for some legal period of time. Such a procedure will be able to capture and analyze all the subject's private data, such as web searches, social media, emails, et cetera....*our very cyber existence while we were alive*...and recreate a cyber persona of each of us. Such data might even be fed into algorithms loaded into AI "personalities" of who we were as individuals to recreate humanoid versions of future generations' *ancestors*! So a family could take in a humanoid version of an ancestor dead a century ago! Talk about a "living" ancestry book!

My little god-granddaughters will very likely have very long lives. They might even be the ones to live as long as the Biblical Methuselah! How can this happen in reality? There are many factors that may extend the average human lifespan, and research with stem cells shows great promise on many fronts, contributing to life extension. The question of people living so much longer then raises the questions of "quality of life," and caring for those who may be elderly for many decades. I hypothesize that people who live centuries will live much of their lives as we do today, so the aging process that is normal today will be "stretched" over the mega-lifespans of the future.

For many reasons, as time moves on, we can expect to have "great-great-great-great" grandchildren and grandparents all living during the same period, because of the increased lifetimes. As lifespans today covers a range, and generations are given "ID's" such as Gen Alpha, decades from now as centenarians become bi-centenarians, then tri-centenarians, obviously a whole new nomenclature of identifying a generation will come about.

Many social and societal issues will arise as people live so much longer. For instance, how will inheritance be handled? When a 678 year old finally dies, should their estate go to children who may be 625, or should the estate go to the youngest family members, who may be 25? One can also ask, should persons over say 75 be allowed to procreate? There are monumental implications to mega-aging that we cannot even begin to fathom in reality, but these issues will have to be, one day.

CHAPTER FIVE – COMPLEXITY & FRAGILITY

Every day, likely multiple times an hour, some business or governmental entity is attacked by hackers. While I was writing this book, ransomware hackers succeeded in shutting down the largest petroleum pipeline in America and a global meat company, until ransoms were paid. There is certainly a preponderance of writing concerning vulnerabilities of our (the world's) data, communications, and power infrastructure and processes, and daily the topic is in the news. Fortunately, there is an industry of people worldwide who are dedicated to leaning forward in assessing the threats to these systems and addressing countermeasures. Thirty years ago, I was one of those people, in the early days of the cyber environment. However, regrettably, those entities that are at risk are not doing nearly enough to protect their assets. In today's supply chain and linked economy, if one major system goes down, it can have widespread effects. We are being warned right now that the entire power grid of the USA could be shut down, and although critical businesses have on-site backup generators, what about the rest of the economy?

I grew up going to the Saturday night picture show to watch mostly Hollywood westerns, and I am reminded of the robbers appearing from over the hill and surrounding and chasing down the stagecoach carrying the box of money or gold bars. So greed and creative attacks are as common today as then, it's just the scenery that is different.

The complexity of the world's cyber systems integrated with every aspect of our lives is already incomprehensible to the average person. In coming years, this complexity will grow exponentially, likely far exceeding the complexities of the human brain and body. And, with increasing complexity comes increasing fragility: meaning vulnerability of systems to threats and attacks from "the outside" as well as internal cascading failures due to some initial error, a single-point failure, or even, lo, a human failure.

Most of you reading this book have at least heard of the "butterfly effect": that a butterfly in say, Vietnam, delicately flapping its wings as it moves from flower to flower, disturbs the air slightly with each beat of its wings, and that disturbance can cause a ripple in the environment that could eventually cause it to rain over, say, Midland, Texas. While challenging to grasp for sure, this analogy has a great deal of truth in it, for Nature is a continuum that our understanding has barely penetrated, perhaps because we are, in fact, a part of Nature. It's hard to see the forest when we are some of the trees.

As time goes by, the coming changes in technology will carry a latent potential to affect the world by influencing all elements of civilization, and thence Nature, in ways we have not witnessed previously. Some of these changes will be akin to the butterfly effect, perhaps a seemingly trivial invention, a new child's toy for example, that becomes so popular that a precious metal used to make it results in the release of more heat to the atmosphere in a particular locale, changing even the local weather pattern, which could negatively affect the growth of a particular type of plant which itself gives off a poison that controls the population of a certain insect......on and on, you see?

The increasing complexity of our world will be so embedded that 99.9% of it will be invisible to all but the "gurus" in various fields. Even they will have a small grasp of the realities of the "cross complexity," which binds all their respective fields together while linking all these to the "mothership," the holistic complexity of the world. The reason for the invisibility will lie with the fact that AI will be "in control" of the preponderance of fields and moment-by-moment activities within every field, so humans will be in many ways "at the mercy" of AI. AI will, of course, evolve with integral and organic self-policing of its logic and decisions, but although AI may be less error-prone than a human, even the slightest error that slips past safeguards could have monumental consequences. None of any such scenarios of error occurrence would be positive, and any such would likely have perturbations throughout the affected system or across a plethora of systems. While errors "can happen," as we know today, due to viruses and hacking of cyber system, tomorrow rogue AI elements "themselves, "…may also become threats. A "rogue" AI element in the overall matrix could seek outcomes beneficial to its true or concocted goals, resulting in harmful outcomes for other fields. Keep in mind that at some point in time, AI will be more intelligent in most, if not all ways, than the human mind. So any AI malfeasance will be nigh impossible to counter.

CHAPTER SIX – FUTURE LOVE, BIRTH AND DEATH

Will love as we know it today evolve (if that's the right word)? Yes, it will. I speak of love for someone else or more than one intimate love, yet Kahlil Gibran, one of the great poets of love, would have to come up with some new verses to describe our future love affairs. Our collective human intellect continually seeks to create the "new" for many and diverse reasons, such as better batteries for electric automobiles, with the intent of reducing carbon emissions. But many innovations derive from motives such as fear or sexual urge, resulting in more inventions that are more powerful, more accurate, smaller, more sensual, et cetera. Sex is powerful motive of the creative….not meaning biological reproduction, but the drive of our sexual imaginations to have "even better" sex and intimate relationships.

For thousands of years, humans have probably found devices to fulfill their sexual pleasure; for example, women making use of certain vegetables as artificial phalluses. We won't get into what men may have used. Certainly, over the last fifty years, there has been an explosion of "sex toys," and thanks to the sexual revolution inspired by my "Love Generation," shopping for such toys or "medical devices" is as easy as searching the internet. Whether its life-like dildos or various anatomically correct "dolls," a man or woman can have it in their bedroom even the next day thanks to the marvels of today's "just in time" (you can laugh!) shipping and delivery systems.

As I mentioned in the Preface, AI and technological advances in robotics and materials have given humankind the capability of very soon being able to create robots that, for all "practical purposes," are fully functional humanoids. These new members of society will each be designed to the human owner's (persons or corporate or governmental) physical specifications, "programmed" with a "starter set" of "personality" traits and behaviors to the owner's desires, and be able to learn and modify its AI based on interactions with its owner, its environment and other humans…and robots (not only other humanoid types but any type of AI robot).

It does not take a great deal of imagination to see and predict that, given the nature of people, the role of these humanoids in our lives will increase exponentially. From being initially compatible "love partners" to men and women who would rather not deal with real human intimacy, robots of various designs and programming will soon begin to permeate our lives in public, not just around the house or bedroom. Already robots serve up drinks in some bars! Will wait-persons, such as the cute young women working at places like "Hooters" or even dancers in "gentlemen's clubs," be replaced with humanoids…who never get tired or need a smoke break outback? One can even envision that far in the future, androids will be going out clubbing just like today's young men and women do! Will "male" androids be seeking the company of cute "female" androids? ("Heh Big Papazoid, how about a lap dance"?) Ok, we will leave it at that.

If I could return to pay a visit to the grown children of my Alpha babies after I leave this realm, I would not be surprised that taking care of their children were humanoid nannies and teachers! It is also likely that these babies were born at home under the supervision of android doulas!

Change in technology and its effects on our daily lives appear to happen rapidly through the eye of hindsight. Still, when we zoom into our day-to-day routines, year after year, the changes are so innocuous that we hardly notice (until we have some gray hair and suddenly blurt out at family get-togethers, "Remember when cars had metal dashboards!").

Our personal and family lives will be so different in twenty-five years from today that even today's Gen Y's will feel ancient. From how people (speaking of those fortunate to live in "First World" nations) prepare their meals, to how our toilet waste is disposed of, even perhaps how we clean our bottoms, will be different from today. From birth to death, the "grandeur" of science and technology will continue to reshape our lives and send waves into the future to propagate more change. How human bodies are "put to rest" even will change. The competition for real estate will eventually preclude expanding or adding more cemeteries, forcing more families to use cremation, and that technology will not be the ovens used today, but "flash ovens" to speed up the process. Or, for those who can afford it, no doubt corpses will be sterilized and then loaded on burial rockets and shot into deep space (at least I hope its deep space and not near-Earth orbits! We already have so much space junk floating around Earth that one day soon no doubt a new business will be salvaging it!)

CHAPTER SEVEN-ROUND, ROUND, GET AROUND

The Beach Boys, 1964...*I Get Around,* reached No. 1 on *Billboard* by July. July found me laying on a Los Angeles beach with my date and my new-found pal at my summer job in downtown LA, and his girlfriend. Oh what a summer! Soon though I would be off to the University of Texas, entering a whole new world of study, eating, beer guzzling, and "getting around"!

We humans have been mobile creatures since we stood upright. To be able to move and travel is in our DNA. First, it was our own two legs, then some of us got smart and figured out we could climb on the backs of horses if we could hang on. Somewhere along the way came the wheel, then carts that horses and oxen could pull. That took care of travel on land for a long while, but those of us who lived by large rivers and seas figured out that those floatable logs could be modified into boats, pulled along with oars or poles, but then the stiff wind that blew against our backs informed us that maybe we could use some animal skins and later on, cloth to create what we call sails.

Thousands of years would pass until the first steam and internal combustion engines were invented, giving means to power what had been pulled by men or horses mechanically. My father's father, a cowboy, told me many stories "back in the day" about long horse buggy rides and traveling the country on horseback. But about that same time Orville and Wilbur were successful at Kitty Hawk, taking civilization a quantum leap into how humans would travel.

We will very soon see another quantum leap in our modes of travel. Hypersonic engines will propel the tubes we sit in comfortably at multiple times the speed of sound as we fly from San Francisco to New Delhi. And, for those business persons who must make a last-minute meeting, there will be the more expensive travel via low orbit rocket planes. These modes of travel are really just a few years ahead. Beyond that likely will be personal vehicles using similar technologies, available to the wealthier, just as self-driving Tesla vehicles are today.

Already the technology exists giving those of us who can afford it, personal driver-less automobiles, fully electric, and we can expect driver-less – or robot or android driven - personal aircraft to whisk us through the skies across town or to another city for a dinner date in the not so distant future. The "Jetsons" will have nothing on us, possibly within my lifetime!

Even now, we see military and others beginning to make use of strap-on "flying machines" that can whisk a person to a destination, then land at a precise point: obviously valuable for military applications. However, we will soon see these used to get emergency medical personnel to accident scenes, or to assist persons trapped on burning buildings, or "you name it." For years we have seen strap-on jet "batman" type suits used by those daring enough, but the use of such technology undoubtedly is not far from being applied in many other valuable applications, where even drones are not the right fit. But think of the future. Within ten years, we likely will see low altitude, crewless drones that will be guided along highways and byways of our large cities, and perhaps even in tourist areas, by AI networks. Calling for an "Uber" today on your cell phone will be replaced with speaking into your embedded wrist device to request an air taxi that will whisk you to your destination and set you down vertically "on a dime," or take you on a guided tour of the Grand Canyon, all in a day.

Of course, if one prefers keeping both feet on terra firma, there will be all sorts of vehicles, four-wheeled, two-wheeled, one-wheel, or running on pneumatic lift – and one day "anti-gravity" drives. These will take the technology that companies like Tesla have on the roads today, amped up exponentially and interfaced with major control systems, to provide coordinated use of routes or prevent collisions even on paths through Central Park. Of course, there will always be free spirits who still prefer old-school skateboards, dodging between the "cyber controlled" vehicles. But wait! One day, there will be androids on skateboards, actually enjoying the experience! (Will they have insurance?)

But what about travel on the oceans and beneath the seas? There is yet so much of our planet that lies out of sight, vast mountain ranges and canyons, with new sorts of species being discovered as we probe more deeply into the darkness. The more we know about the deep sea, the better, in order to literally save our seas! Moreover, a day may come because of climate change that humankind must retreat to undersea cities!

Many of us of my generation remember the adventures of the Nautilus, commanded by the genius, Captain Nemo. *20,000 Leagues Under the Sea* – released by Disney way back in 1954, fully captured my eight-year old imagination. I remember playing the role of Captain Nemo out in the woods of the ranch where we lived. There was a giant, ancient fallen oak trunk that had lost all of its bark, and was bleached white, and in my mind it was both a whale and a ship and a submarine. One day, I expect within the next fifty years, we will travel and explore the depths as tourists, and much later begin to construct expansive habitats far beneath the waves.

CHAPTER EIGHT – WORK & LEISURE

Modern humans love their pets. Pets, especially dogs, make up a significant part of many of our lives, whether while sharing our leisure time with these furry friends, or for some people, our work activities. When I was a boy living in the country, we had a feisty little mixed dog, Amigo, who was my best friend until a rancher in a pickup truck ran him over. Amigo had a fatal flaw. When I was away at school, not around to pal with him, he would wander off. One day he found a newly born lamb and decided to make it a meal. That day was his last when he was observed on the highway by the passing rancher. We lived in sheep country, and were sheep and goat ranchers. I suppose my father was likely spared the awful "deed" of eventually having to shoot Amigo. Back then there were not any SPCA's, at least in our ranch country. Later in life, my late wife and I had our dog, Sassy - Sarah Vaughan's nickname (my wife was a well-known jazz singer, so anything personal with her had to have a jazz vibe!). We also had several cats – Fuzz, the matriarch of our clan, being the most memorable over many years. I am one of those "cat and dog lovers."

As humans "evolve," will our pets also be subjects of our advanced sciences? I expect so: hell, all breeds today are creations of mankind's (I mean "man," for women can't be blamed for creating hunting, guard, and fighting dogs, et cetera). Just look at the spectrum of dog breeds: shapes, sizes, disposition, appearance…..all due to man's tinkering with the breeding of those four-legged creatures that first hung around the camps of early humans.

Will our new pets be replaced by android pets that are as "real" as your dog Spot, and much cuter than the little toy mechanical dogs that have been around for years. *Dognoids*? Dogs that don't have to be walked and cleaned up after? Dogs that are already trained and just need to be "fine-tuned" – right out of the box! (And, *catnoids, hampsternoids, turtlenoids,*….all our favorite pets will come to "life" as replicas of nature.)

And, what will be the effect of our technical evolution on canine blood "sports" taking place in the backlots of cities and small towns? Will it become more "attractive" for such cruel misfits to instead buy android dogs to fight? It seems to be in our DNA to want to see battles for life and death, whether in the Colosseum of Rome or as readily evident in today's popularity of robot battles and little boys' fascination with fighting dinosaur toys. But as I ask elsewhere, given that the same technology that will create life-like android "humans" can be applied to any other species, robot dogs and other animals may come to feel pain as profoundly as real animals, so will our morals and ethics keep stop us from inhibiting the AI sensations of pain and the accompanying feelings? I pray that by then the SPCA expands its umbrella of protection!

If someone can fall in love with an android partner, then we can expect people to bond with robot dogs and other species, perhaps. I can see many benefits to this, besides not having to clean up after Pooch, or worry about what to do with this member of the family when the humans are away on vacay. First, it would reduce the population of real dogs exponentially, thereby reducing the demand for raising other animals to provide them feed, which in turn contributes to reducing emissions into the atmosphere and chemicals into our waters. It would also remove from people's lives the heartbreak of having to euthanize their dogs, for dognoids would never get sick and would never die!

Aside from the furry species in our lives, one can think of a million ways in which our work and leisure will change (as with other future changes, I hesitate to apply the term "evolve"). But let's delve into the potpourri of "could happen."

As someone who loves music, especially jazz, old school R&B, world music with a beat and certain film genres, I can appreciate the possibilities in the music and film industry of the future. Presently virtual reality systems and interfaces are well developed, indicating that future advances in this area will foster a new world for participants and audiences.

At some point in the not too distant future, theatergoers will each don some type of VR (virtual reality) headset built into the theater chairs and become active participants in the movie. With the power of supercomputers of the future, likely using quantum computational structures, the power of VR will be unlimited. Of course, those of the future should expect that such technology will be readily available at home, whether one wishes to engage in a geographical expedition, or enjoy a *ménage à trois*.

Today DIY music creation technology allows one who is talented, but not inclined to study music formally, to learn how and produce music digitally. Synthesizers have been around for decades now, but with AI overlaid on the concepts of these systems, even amateurs will be able to create scores of any style of composition and expression. So what then will be the future of the music business? Will there still be copyrights since the capacity to produce "designer" music will be in the hands of about anyone with extra money to spend and with the talent to use it? I have this vision that at some point, the creation of music and visual imagery will meld, somewhat like incidental music of today's films, but to a holistic degree that allows creators to submerge the audience in incredible experiences.

Since Star Trek hit our teles, we have seen the usefulness of what were called "holodecks." While one cannot grasp how those systems "actually worked", technology and science clearly point to a future where such virtual reality spaces will be possible. But it is not a stretch at all, even from the technology that exists today for holographic projections, to see the first step into the future will allow the viewer to step into the holographic experience. Far into the future – but not that far – we will be able to couple our brain directly into the experience, much like what is depicted in the film "Pandora."

Speaking of brain interfaces, I do not doubt that at some point in time, again not very far into the future, people will be able to link into various sorts of AI "games" or other experiences and participate with other humans remotely in all kinds of situations; including, obviously, sex. One cannot but wonder what will become of the "sex worker" trade as "non-reality" becomes "reality," populated with humanoids and AI sex potpourri. Even today in

Japan there is a problem with young people dating and having sex, and making babies, because so many of the young men, and girls, are hooked on porn. Likely in the future the population growth of developed nations will take a down turn for similar reasons.

We will discuss "future work" below, but let's continue looking into what leisure might look like in the future. First, travel times anywhere will be greatly reduced, thus leaving more time for experiencing the destinations. Destinations of the future will be both virtual and real, and real will include nature herself, as well as "nature like creations" – think of Disneyworld on mega-steroids. We can expect the bartenders and wait staff to be humanoids at resorts, who laugh and smile tirelessly at the rudest tourists. Lifeguards will also be some type of robot, able to swim through any surf to bring aid or rescue. Guides for climbs up the highest mountains may be androids as well, able to traverse the worst ice and be totally resistant to the cold, assuring the safety of climbers. Rescues on Mt. Everest will be done using androids, perhaps using "jet packs" and intelligent drones.

What will become of today's sports? I believe that such sports will remain intact, pretty much "as is," although we can expect stadiums and facilities to evolve. Humans will find an increasing need to "default" to human-to-human contact and experience in such a hyper-technology world. This in itself could lead to a bifurcation of humans and androids far into the future, as both seek comfort and complete empathy from those of their ilk. However, we also will see the new sports appear, where humans and humanoids participate. And, returning to that theme of how many people thrill at violence, e.g., the mixed-martial-arts of today, just think of what may return: a Colosseum of fighting humanoids! I shudder at the vision of this.

However, we can also expect new forms of sports to come into the mainstream, just like pole dancing (now a sport), ski boards, and skateboards have! Many people will continue to seek the rush of adrenalin, using new technologies to do more and more risky activities. There seems to be no limit to our creativity when it comes to getting "the rush." We will also have at our leisure living room games, including connections that will hype our sensory systems to enhance the feelings of danger and "the rush."

Now we come to the flipside of leisure that being our future work. All the speculation on what leisure will be in the near and very far future informs us also about the changes to expect in our work, our work environments, and the value of the work experience to our person, our social life and our societies.

We already can see the impact of robotics and AI within the industrial workplace. In the enormous warehouses and distribution centers for the "big box" and online shopping "stores," robots have been hard at work for decades, actually. However, the average consumer is oblivious to this, being content to receive what they ordered while sitting comfortably at their computer or by using their smartphone. But these same people might be shocked to find out that the reason their uncle's career took a turn for the worse was largely due to the company he had worked for many years made a decision, based on the direction of the competitive, profit-driven free market was taking, to "retool" their entire delivery strategy by mechanizing and robotizing a large part of their operations. Uncle Joe was left out in the cold one Friday, wondering what he could do for the remainder of his working life.

In most cases, men like Joe, and women like Sandra, were too old or did not have the personal finances to choose to go back to school or college to learn a new trade, such as robotic maintenance or programming. So like thousands of others, they fell into the growing class of perennially under/unemployed, teetering on falling into the "lower class," slogging along among the lower middle class of Americans. (I limit this discussion to the USA, as various other developed countries have responded differently to industrial advances in the workplace.)

Presently, there is a clear and urgent need for educational and social responses to what will be a continuing and growing phenomenon: the increasing conversion of basic work tasks from humans to machines. Increasingly, we hear calls for college and trade schools to target industry-specific needs, and it is evident that more and more of such training curricula are becoming available. However, the industry needs greater participation in developing such programs and providing benefits to assist employees in retraining. Until the mentality of hiring the younger, less expensive worker over retraining the veteran employee changes, the USA will be fighting a competitive battle with other nations such as China, with one hand tied behind us. In time all workers will be those working within human/humanoid "labor" unions.

When this era opens, humans will be trained in working with the high-level aspects of operations and maintenance of these systems, and jobs for people then will range from that of "watchdog" roles monitoring the highest-level of AI activities to actually interacting with humanoids as "cyber-psychologists" dealing with the feelings of humanoids about their roles and jobs.

Compared to today, jobs for people will be reduced, and the "work week" will be radically reduced, with humans likely "working" almost entirely remotely. This translates to opportunities to have increased quality of personal and family life, and of course, leisure time. Some European countries have long valued shorter work weeks, longer vacations, and overall mitigating stress. So perhaps here in America, we will find that ideal situation in the not too distant future, and beyond that ratio of work hours to leisure hours will only grow.

One can ask, however, if robots and AI take jobs from humans, where will humans fit into the mix of our economy? Assuredly there will continue to be an increase in the so-called cottage or mom & pop businesses. Still, these will be riding the technology wave also, providing services that the large companies choose not to compete in. However, although politics does not deal with this well, one thing that must happen is the educating and training of new low-skilled immigrants, especially those undocumented persons and families seeking refuge here.

Legal and undocumented migration into the USA will increase more and more over time due to the effects of climate change on the poorest areas of the planet. It is imperative, beginning now, for our society to fully recognize the present and coming plight of low-skilled workers, for instance, those in the hospitality and food industry. There are a myriad of factors impacting these sectors of our economy, but prospects over the coming decades will continue to dim as robotics and AI are dispersed throughout all types of businesses. The continuing education of lower-skilled workers will have to be responsive to these changes, mitigating an explosion of displaced workers losing out, becoming homeless, or turning to crime.

We can expect the daily lives of humans to be "light years" different from how we live today. The experiences of how we shop today, whether "online" or going to malls or corner stores, will become obsolete. At some point, we will be able to design whatever we want in the way of typical consumer items, clothes, et cetera, through AI, and we will be able to make many of these rights at home or "office," using "3-D printers," except more akin to the "teleporters" of Star Trek films. Holographic imaging will allow us to design and "try on" clothing, then say one word, "Make," and within seconds or a few minutes, we will have a new dress, suit, or shoes. Beyond that will come a time when we step into a "booth" and whatever we are wearing for that day and activity will be applied to our bodies….that is, we will be dressed in designs and materials which are formed by swarming nanobots, and our clothing will be holistically compatible with our needs of that particular day.

Many other experiences will be via holographic reality, physically immersing us in situations such as conferences, games, sports, sexual activity, and various forms of competition. People are living today who always say they would like to go to a nude beach….well, live long enough, and you will be able to without even shedding your real clothes!

CHAPTER NINE– SPACE IS THE PLACE

One of my favorite jazz artists, Sun Ra, said so. His 1974 film by that name is a monument to Afro-futurism. In that iconic, eclectic film, Sun Ra, as an enlightened being from another dimension, comes to Earth to instruct African Americans and open their eyes to the reality of racist America. (As a side note, I had the pleasure of meeting Sun Ra and a few of his Arkestra once; he was definitely human, and a very nice fellow!)

While there are probably limitless interpretations of the messages Sun Ra painted in his music and visual iconography, a theme is evident that is universal to humankind. Humans have looked to the heavens since we began to walk erect as early hominids in East Africa. Embedded within our consciousness is the gnawing question of "what is out there?" Of course, "out there" began really as "what is over the hill?", or "is there anything beyond the horizon of the sea?" We are explorers and in various cultures across time, this predisposition has expressed physically, mentally, and spiritually.

In 1959 when JFK declared that we would walk on the Moon by the end of the coming decade, the chart into the heavens began, although the course would be adjusted as each decade rolled by. Today NASA and private firms are looking toward a human mission to Mars, following up on the now decades of exploration of that planet by the Rovers and recently, the little helicopter, Ingenuity. Yet one can legitimately inquire, "why should we send people to Mars, or beyond, when we seem to be at a tipping point on saving our planet from the curse of climate change"?

Regardless of how one feels about this thrust into our solar system, that we will continue to push further into space seems predestined. The timing and nature of crewed missions to Mars will be the next chapter in our exploration. Presently in 2021, the myriad of challenges in creating living accommodations on the Red Planet for us fragile humans and providing for water, food, and other necessities for long-term habitation, have to be overcome. However, the most significant challenge at this point, unless these pioneers are committed to a one-way trip, is how to get off the planet; i.e., where do the astronauts get the fuel for their rocket? But that too is being diligently addressed.

I have no doubt we will figure out solutions to all these issues. We always have as human explorers here on Earth and off-Earth. But as futurists, we must also consider not "just" the space travel and living on Mars, but also what technology and scientific discoveries will spin-off from this push. So many things we take for granted today did not exist when I was a boy, and what we enjoy now were spun off….at least the original technology was…from the space program of the '60's decade. As an aside, there is a more profound irony, or perhaps a truth, if we roll the film back to the days right after the end of WW2. At that time there was a scramble by both the United States and Russia to acquire the secret technologies of Hitler's regime. The Germans were far ahead of the Allies in military technology, so Americans and Russians both sought to obtain as much of that pie as they could. The USA was able to round up dozens of scientists and engineers, including Werner Von Braun and colleagues. Yes, they were Nazis, who had used their brains to create new inventions in a futile attempt to crush the Allied and Russian advances. However, that corralling of these brilliant engineering minds gave the USA a jump start in its rocket program.

Over the 1950's that activity continued during the Cold War with Russia. From that program came the rockets that would take us to the Moon. The irony I mention is that out of a time, and people, responsible for the Nazi horrors came a technology that over the years flowed into our daily lives, improving the world. Since Neil Armstrong took that first giant step for mankind, the continuing space program can be credited directly and indirectly for the preponderance of "things" that we enjoy today that make our lives better in countless ways. All of this happened in the last 50-60 years: it is impossible to forecast what space travel and technology will yield in return in the next half-century, but undoubtedly it will be beyond our imaginations!

Yet that is what futurists do: shed the shackles of being unimaginative, and dream the future! So let's close our eyes and let our minds roam free. Space travel will just be "travel" in a very few decades. Already within just a few short years, travel in near orbit altitudes will be commonplace, with trips that now take many hours being traversed in a mere fraction of that via rockets that will take off and land just like airliners have been doing for a 100 years now- except the landing will be vertically, as is already being done by SpaceX. As an engineer who has experience designing large airports, I can see that airports to accommodate such passenger rocketry will be very different.

But as exciting space travel is for humans, one must ask whether, with the advances in robotics and the current rate of development of AI and robots, does it make sense for humans to create any large colony on Mars or even consider going further into our solar system. Why not "ask" our creations, computers like "HAL" (in the 1968 film, *2001: A Space Odyssey*) and robots that may or may not be anatomically "human," to do the traveling for us? Most likely, it will be some combination of people and AI/Robots, likely with the first very deep explorations being done by robots. By robots, I do not mean machines like the latest mission to Mars, but rather robots that are designed for specific functions, not necessarily humanoid in appearance, for such flattering of our species will be superfluous for such endeavors. But also, like "HAL," the AI will be integral to the very structure and systems of the space ships. I can also see that in time we likely will genetically engineer humans for space travel. By then, our "civilized" values will have changed to accommodate the notion of modifying certain human embryos to be destined for deep space travel. Such human lines will not necessarily even be born on Earth, but maybe placed aboard space ships that will travel for many years, during which these humans will be born from artificial uteruses and raised by robot nannies, taught what they need to know about Earth, the human species, and what their mission in life is.

CHAPTER TEN – END OF RELIGION

Will we see the death of "God" one day (or have we already"?). The end of world religions? One may speculate that the days are numbered for beliefs in a Supreme Being and the world religions created around such notions of a Deity and Creation. As the mass of humanity becomes more educated through access to knowledge, and more and more revelations point to science over creationism, will reason prevail? Will people come to understand that the kingdom of God truly is within oneself, as stated in Luke 17:21, and as in the title of Leo Tolstoy's book? Will most people finally grasp that the great books of world religions – and "unwritten books" of indigenous spirituality – are (as I believe) fragmented spasms of history wrapped in mythology? Religions will not entirely fade away, but rather that they will mutate, like viruses, yet maintain at least some thread of connection to their origins (I have used the term "viruses" because it is my firm belief that most religions have had a negative and harmful effect on believers and civilization. Religions were all created by flesh and blood *male* humans to control the masses by turning myth into scriptural texts, giving people the false notion – a crutch – to make life tolerable, and to allow a priesthood to control the masses and their wealth. Maybe one day people will realize that "God" is only within oneself, particularly one's mind, which is but less than fleeting spark in the cosmic Universal Consciousness.

Today, only a few physicists believe in a "divine God," Generally, surveys indicate that only about 4 out of 10 scientists believe in God. This contrasts with over 9 out of 10 average Americans who profess a belief in God or a higher spirit. Even a number of well-known Biblical scholars/archeologists, raised as Christians, having spent lifetimes digging in the Middle-East trying to connect the pieces of Biblical stories to real history records, admit they are no longer "believers" in the conventional sense.

Of course, how one defines "divine God" plays into the discussion significantly. Are we speaking of God of the Bible, Allah of the Koran, the many aspects of God of Hinduism, or Buddha Shakyamuni, or as manifested in indigenous religions? By my way of thinking, if one wishes to acknowledge "God," the gods of all religions are the same.

Or, do we mean some infinite, unknown source of intelligence that laid out the minute plans for creating our universe – and all the other possible universes that modern-day physics postulates (i.e., parallel universe, multi-universe, et cetera). Personally, I opt for abandoning the human notion that everything must have a beginning. I subscribe to the theory that the creation of one universe is brought about by the collapse of another, in such an event as the "Big Bang." There is no creator needed in this theory, but since we cannot grasp "infinity" (as a chain of events with no beginning nor end), we humans generally are left puzzled. But "who" invented the beauty of this never-ending cycle? I contend it is only the present incipient, developmental stage of our human minds that cannot comprehend infinity, so the easy default is to seek a "God."

However, religion in general will have to come to terms with questions that the future will bring, such as, does a cloned human have a soul? Does a humanoid robot, with feelings and a sense of the "self," have a soul? (So, do we become "God" to these new creatures? Will new religions spring forth, where humans write "the good books" for humanoids to abide by?)

I would be remiss, I believe, in not mentioning at this juncture that as I write this book, there is a serious concern about the reality of UFOs – unidentified flying objects. So, what if we Earthlings do find out that we are being watched by beings from another world (our universe), or are at last directly contacted by these beings? Questions arise in my mind like, do they have a religion? Is their "God" also our "God," and how do they think of "God"? We could go on and on along this line of query, such as "Do they have their own Jesus"? "Do we share the same heaven or hell as they"?(I will leave it at that.)

As mentioned earlier, there will come a time, I believe relatively soon, certainly within a few decades, when humans will bond romantically with android (humanoid) partners, and our laws in the modern nations of that day will begin to allow the marriage of people and androids. But then what about offspring from such marriages? At first, these will be done using some form of surrogate human females, but later artificial uteruses will nurture and give life to the newborn. Somewhere along the way, I can see some "DNA" imprints of the android partner being "downloaded" from its brain into the new child. By this time traditions such as Christening and various other ceremonies associated with the newborn will be the lore of ancient religions.

We humans will also be tempted, and no doubt bite the apple in areas such as advanced gene modification tools, going far beyond today's genetic engineering. We will be playing with Nature herself, "fixing" things we do not want or like. This may be gene editing of human creation, or cloning of humans for specific purposes, or playing with Nature at large as we dispatch various potions into the natural environment to modify or destroy certain viruses and bacteria (not knowing what the repercussions will be). Do we become our own "God" at that point? I shudder to contemplate the return of some "advanced" form of "eugenics" that could result in the "haves" attempting to build their own superior "race." Where have we seen that before?

The arrow of time will take humankind into vast uncharted territories of ethics, morality and what religion means. My future reincarnations look forward to a front row seat to it all!

EPILOGUE

My father was a B-25 bomber/command pilot in WW2. On VE day, May 8, 1945, he received the Distinguished Flying Cross. Less than a year later I came into this world. It is primarily from a slice in time from 1946 up to today in 2021 that my perspective in writing this futurist book sprang forth. My own days in this world are, shall we say, numbered, yet my two god-granddaughters may easily see another 100 years, and I believe, as mentioned above, very, very much longer.

Given the pace of discovery in science and its applications in technology, I expect marvelous changes will be witnessed within the next 10-20 years. Beyond that, the sky is *not* the limit. However, I suppose that until AI becomes so intelligent as to begin to determine its own future, humankind will continue creating a mix of "new things." Many of these "things" will come forth out of necessity for the survival of our species, and hopefully much of the remaining flora and fauna of our Mother Earth. Many other creations and inventions will also come along to satisfy our human need for fun, ease of work and living, and "just because."

Was the "apple" Eve offered Adam as told in the Bible, a metaphor for consciousness? With a bite from the apple, Adam gained human consciousness. He became self-aware, eventually gaining a conscience, and with that awareness indeed came the question that we humans have borne through the ages, "Is there more?" That eternal question provides the motive for our human hunger for knowledge and our quests, and will continue to do so throughout the coming ages.

Implicitly embedded in the entirety of this book are the societal issues of justice and the rule of law. Justice and the rules of law will change over the lifetime of Gen Alpha, and radically so beyond. Today laws are being "made" and "unmade" and revised almost daily somewhere in the United States. Laws are not blind but are extensions of the values of those who have control of law-making at any point in time, whether it be the fragile democracy we enjoy here, or the laws set forth by Sharia in some Islamic countries, or the law laid down by dictators. I would like to believe that as decades go by, the fairness of laws increases. But one can, of course, ask what the definition of "fairness" is. With the coming of AI, one day, we may see juries using AI to fully examine the evidence they heard during the trial, and perhaps even the judge will be an android. Perhaps then the verdicts will be truly based on the evidence, devoid of prejudices of law enforcement testimonies, prosecutor bias, and jury members' prejudices.

And what of law enforcement? Given the crisis in policing in the USA today, with repeated incidents of cops acting as lynch mobs, mainly toward people of color, an overhaul of policing is much needed. Someday, far in the future, I do not doubt that cops will mostly be androids or other forms of robots. We have already seen such scenes in various sci-fi films, such as robot cops riding "motorcycles" that fly. We can hope that these cops will have learned the proper ways to best handle human crises.

Oh, politics (or "politrix," as some Jamaicans say)! Today in the United States, we are witnessing our politics be consumed by conspiracy theories and other untruths, to the point that one major party has now become a true cult by any definition. Democracy itself today is in peril, and at the root of this is white racism. I naively believed for many years that my father's heroism in WW2 as a B-25 bomber commander and his 70 missions over Italy bombing railroad bridges and Nazis was to do away with such hate as promulgated by Hitler and his ilk. But for the last 50 years of my life, I have seen that white racism was as alive in America as in Germany, and lies at the core of the "Big Lie." However, as I alluded to at the start of this book, I choose to believe that over time racism will greatly fade, unfortunately to be replaced with greater classism chasms between the "haves" and "have-nots."

All of this may also translate into major modifications to our Constitution, to the point where cities and clusters of cities having great affinity on values become more important than States and state rights. Will major cities/city groups become "city-states," akin to ancient Greek, Roman, African, and Olmec cultures? I believe that may very well happen because people will tend to align themselves physically by economic class, just as we always have. Because of the power of cyber communications and AI, there will be a "natural" tendency for people to cluster around those affinities that are most attractive to them.

I can also see that far into the future, our northern and southern borders will vanish for all practical purposes. America of the distant future will take on a new name – "North America," and will include all of Central America and Canada. The government will be akin to that of today's EU. The driving forces for such a creation will be the utter failure of Mexico, the great resources south of the USA border, and the opportunities for growth across Canada.

Looking at the spectrum of potential/possible changes that will confront our world over decades and the next century and beyond, I struggle to keep my usual "glass half full" view. After all that we of the human collection have gone through, especially during my lifetime so far, I yearn to feel that our humanity has grown as a result. Yet when I see millions of white Americans falling for the "Big Lie," my optimism withers. However, when I put on my "long lenses" and look at where we have been….Nazi Germany, the dropping of "Fat Man" and "Little Boy" on Japan, Jim Crow, and 4000-plus lynchings and innumerable "race riots," the Vietnam War, 9-11 and the subsequent wars that continue even today, and our continued onslaught against Mother Nature….I see that the history of humanity is replete with lows, yet we continue to pull our pants up and move along. Survival is built into our DNA, and in time all these survival experiences surely will result in collective learning that there is a better way.

As was stated in the Introduction, we are at a "tipping point." We cannot directly paint what the world will be like in 50 or 100 or 150 years, but we can paint, in broad strokes, the world of 5 and 10 years from now. These near-horizon works will appear to be abstract paintings, no doubt, given the complexities at hand and the vicissitudes of human nature and its political expression. The interpretation of these abstracts in time will be indicators of whether we have tipped or whether we have found some stable ground. For the sake of my two little Alpha's and their future progeny, I pray humankind gets it right.

While writing this book, the United States government revealed information about UFOs (unidentified flying objects…commonly associated by UFO enthusiasts with "aliens" from "beyond." This information is not "leftovers" from the Roswell, New Mexico "aliens" of the 1950's, but personal sightings by trained military pilots' eyes and via military imaging systems that have confirmed the *possible* reality of craft with capabilities generations ahead of anything known by US intelligence. Our government does not know what these "UAP's" ("unidentified aerial phenomena," in military-speak) are. So everything in this book could be said to be cast in the light of what this could mean. Personally, I do not believe these sightings are beings from another star system, nor probes from such advanced beings, but rather a collection of various types of phenomena. If I am proven wrong, then actually, I will be thrilled! If I *am* proven wrong, our entire future may be directly or indirectly influenced by these visitors. Or, we may never know for sure, if such beings are real and choose to remain a mystery.

Having read this book, I hope that the reader will take it upon her/himself to dig deeper into the issues at the horizon of the future. These are exciting times to live in, and the future appears dazzling!

www.ingramcontent.com/pod-product-compliance
Lightning Source LLC
Chambersburg PA
CBHW070815220526
45466CB00002B/674